INTRODUCTION

Quantum jumping, also known as reality shifting or dimension jumping, is a metaphysical concept that involves the intentional or unintentional movement of an individual's consciousness between different realities or parallel universes. The concept is based on principles of quantum mechanics, a field of physics that studies the behavior of matter and energy on a subatomic level.

Quantum jumping is a popular topic in the new age community, with many people claiming to have experienced it themselves. However, the concept is often met with skepticism and criticism from the scientific community, as there is little empirical evidence to support its claims.

In this article, we will explore the fundamentals of quantum jumping, its different types, techniques, benefits, challenges, and the ethics involved. We will also examine some real-life examples of people who claim to have experienced quantum jumping and discuss the potential applications and future possibilities of this phenomenon.

Understanding Quantum Mechanics

To understand the concept of quantum jumping, we must first understand the basics of quantum mechanics. Quantum mechanics is a branch of physics that studies the behavior of matter and energy at the subatomic level. It is a relatively new field, with its origins dating back to the early 20th century.

One of the fundamental principles of quantum mechanics is the concept of superposition. Superposition refers to the ability of particles to exist in multiple states simultaneously. For example, an electron can exist in multiple positions at the same time.

Another important principle of quantum mechanics is the concept of entanglement. Entanglement occurs when two particles become linked in such a way that their properties are dependent on each other, regardless of the distance between them. This concept has been used to develop quantum computing and quantum communication.

Basic Principles of Quantum Jumping

The concept of quantum jumping is based on two key principles of quantum mechanics: superposition and entanglement. According to this theory, an individual's consciousness can exist in multiple states or realities simultaneously. Each reality represents a different version of the individual's life, with different outcomes, events, and possibilities.

Quantum jumping occurs when an individual's consciousness shifts from one reality to another. This can happen intentionally, through specific techniques and practices, or unintentionally, through spontaneous events or experiences.

The theory suggests that every decision we make creates a new reality. For example, if we decide to take a different route to work, this creates a new reality where we did not take the usual route. Each reality exists in parallel to our current reality, and we are constantly shifting between them, even if we are not aware of it.

Schrödinger's Cat Thought Experiment

One of the most famous thought experiments in quantum mechanics is Schrödinger's Cat. The experiment involves a cat in a sealed box with a vial of poison. The poison is released if a radioactive atom decays, which is an event with a 50/50 probability.

According to quantum mechanics, the atom exists in a state

of superposition, where it is both decayed and not decayed at the same time. This means that the cat is both alive and dead simultaneously, until the box is opened and the state is observed.

The thought experiment illustrates the concept of superposition and the idea that reality is dependent on observation. It also raises questions about the nature of reality and the role of consciousness in shaping it.

Uncertainty Principle

Another important principle of quantum mechanics is the uncertainty principle, which states that the position and momentum of a particle cannot be measured simultaneously with perfect accuracy. The act of measuring one property affects the measurement of the other property.

This principle has been used to explain the role of consciousness in quantum jumping. According to some theories, the act of observing a particular reality collapses the wave function and determines which reality becomes the observed reality.

The concept of quantum jumping, also known as reality shifting or dimension jumping, is a relatively new concept in metaphysics, but it is based on principles of quantum mechanics, a field of physics that studies the behavior of matter and energy on a subatomic level. In this article, we will explore the history of quantum jumping, from its early origins to its current status as a popular topic in the new age community.

Origins of Quantum Mechanics

The foundations of quantum mechanics were laid in the early 20th century, with the work of physicists such as Max Planck, Albert Einstein, Niels Bohr, and Erwin Schrödinger. Their work challenged the classical view of physics, which held that matter and energy behaved in predictable ways, and introduced new concepts such as wave-particle duality, superposition, and uncertainty.

In 1927, Werner Heisenberg introduced the uncertainty principle, which states that the position and momentum of a particle cannot be measured simultaneously with perfect accuracy. This principle had profound implications for the study of quantum mechanics and sparked a new era of research into the behavior of matter and energy on a subatomic level.

Quantum Mechanics and Consciousness

The study of quantum mechanics has led to many philosophical and metaphysical debates, including the role of consciousness in the universe. Some physicists and philosophers, such as Eugene Wigner and John von Neumann, have argued that the act of observation or measurement collapses the wave function and determines the outcome of a quantum event.

This idea has been used to explain the concept of quantum jumping, where an individual's consciousness can move between different realities or parallel universes. The idea of multiple realities is based on the principles of superposition and entanglement, which suggest that particles can exist in multiple states simultaneously and become linked in such a way that their properties are dependent on each other.

Early Origins of Quantum Jumping

The concept of quantum jumping can be traced back to the early 20th century, when physicists such as Max Planck and Albert Einstein began to challenge the classical view of physics. In 1935, Erwin Schrödinger proposed his famous thought experiment, Schrödinger's Cat, which illustrated the concept of superposition and the idea that reality is dependent on observation.

The concept of parallel universes was also explored by Hugh Everett in 1957, who proposed the Many-Worlds Interpretation of quantum mechanics. This interpretation suggests that every time a quantum event occurs, the universe splits into multiple parallel universes, each containing a different outcome.

However, it wasn't until the 1990s that the concept of quantum jumping began to gain popularity in the new age community, with the publication of books such as "Reality Shifts" by Cynthia Sue Larson and "Quantum Jumps" by Cynthia Sue Larson.

Popularity of Quantum Jumping

In recent years, quantum jumping has become a popular topic in the new age community, with many people claiming to have experienced it themselves. The concept has been popularized through books, blogs, social media, and online forums, where individuals share their experiences and techniques for intentional reality shifting.

The popularity of quantum jumping can be attributed to several factors, including the growing interest in metaphysical and spiritual practices, the increasing availability of information and resources online, and the desire for personal growth and transformation.

Techniques and Practices

There are many techniques and practices that are used to facilitate quantum jumping, including visualization, meditation, affirmations, quantum jumping exercises, and quantum jumping journaling. These techniques are designed to help individuals shift their consciousness from one reality to another, either intentionally or unintentionally.

Visualization is one of the most popular techniques for quantum jumping, where individuals visualize a different reality and focus their intention on moving into that reality. Meditation is also commonly

Quantum jumping, also known as reality shifting or dimension jumping, is a concept that has gained popularity in recent years in the new age community. It is based on the principles of quantum mechanics, a field of physics that studies the behavior of matter and energy on a subatomic level. In this article, we will explore the importance of quantum jumping and its potential benefits for personal growth and transformation.

The Importance of Quantum Mechanics

Quantum mechanics has had a profound impact on our understanding of the universe and has led to many technological advances, including the development of transistors, lasers, and MRI machines. It has also challenged our classical view of physics, which held that matter and energy behaved in predictable ways.

The study of quantum mechanics has led to many philosophical and metaphysical debates, including the role of consciousness in the universe. Some physicists and philosophers have argued that the act of observation or measurement collapses the wave function and determines the outcome of a quantum event. This idea has been used to explain the concept of quantum jumping, where an individual's consciousness can move between different realities or parallel universes.

The Importance of Personal Growth and Transformation

Personal growth and transformation are important aspects of human life, as they allow us to become more self-aware, develop new skills and abilities, and achieve our goals and aspirations. Quantum jumping can be seen as a tool for personal growth and transformation, as it allows individuals to shift their consciousness from one reality to another and explore new possibilities.

Overcome limiting beliefs and patterns of behavior

Many individuals have limiting beliefs and patterns of behavior that prevent them from achieving their goals and living their best life. These limiting beliefs may be based on past experiences, societal conditioning, or negative self-talk. Quantum jumping can help individuals to overcome these limiting beliefs and patterns of behavior by allowing them to shift their consciousness to a reality where these beliefs and patterns do not exist.

Access new possibilities and opportunities

Quantum jumping can also help individuals to access new possibilities and opportunities that may not have been available to them in their current reality. By shifting their consciousness to a reality where these possibilities and opportunities exist, individuals can gain new insights and ideas that can help them to achieve their goals and aspirations.

Enhance creativity and problem-solving skills:
Quantum jumping can also enhance creativity and problem-solving skills by allowing individuals to access new perspectives and ideas. By shifting their consciousness to a reality where these perspectives and ideas exist, individuals can gain new insights and inspiration that can help them to solve problems and come up with new solutions.

Improve mental and emotional well-being:
Quantum jumping can also improve mental and emotional well-being by allowing individuals to shift their consciousness to a reality where they feel more empowered, happy, and fulfilled. This can help to reduce stress, anxiety, and depression and improve overall quality of life.

Quantum Jumping Techniques

There are many techniques and practices that are used to facilitate quantum jumping, including visualization, meditation, affirmations, quantum jumping exercises, and quantum jumping journaling.

Visualization is one of the most popular techniques for quantum jumping, where individuals visualize a different reality and focus their intention on moving into that reality. Meditation is also commonly used to facilitate quantum jumping, as it allows individuals to quiet their mind and focus their intention on the desired outcome.

Affirmations are another powerful tool for quantum jumping, as they allow individuals to program their subconscious mind with positive beliefs and affirmations. Quantum jumping exercises and journaling can also be used to facilitate the process, as they allow individuals to track their progress and gain new insights and ideas.

Quantum mechanics is a branch of physics that deals with the behavior of matter and energy on a subatomic level. It is a fundamental theory that has revolutionized our understanding of the universe and has led to many technological advances, including the development of transistors, lasers, and MRI machines. In this article, we will explore the fundamentals of quantum mechanics, including wave-particle duality, the uncertainty principle, and the Schrödinger equation.

Wave-Particle Duality

One of the fundamental principles of quantum mechanics is the concept of wave-particle duality, which states that particles can exhibit both wave-like and particle-like behavior depending on how they are observed. This principle is based on the fact that particles, such as electrons and photons, have both wave-like and particle-like properties.

In a classic experiment, called the double-slit experiment, electrons were fired through two slits and onto a screen behind them. When observed, the electrons behaved like particles, forming a pattern of two lines on the screen. However, when the experiment was repeated without an observer, the electrons behaved like waves, forming an interference pattern of multiple lines on the screen.

This experiment demonstrates the concept of wave-particle duality, where the behavior of particles depends on how they are observed. It also highlights the fundamental nature of uncertainty in quantum mechanics, as the behavior of particles cannot be predicted with absolute certainty.

The Uncertainty Principle

The uncertainty principle is another fundamental principle of quantum mechanics, which states that the position and momentum of a particle cannot be measured simultaneously with absolute accuracy. This principle is based on the fact that the act of measuring one property of a particle affects the other property.

For example, if the position of a particle is measured with absolute accuracy, its momentum cannot be measured with absolute accuracy. This is because the act of measuring the position of the particle affects its momentum.

The uncertainty principle places a fundamental limit on our ability to predict the behavior of particles in quantum mechanics. It also highlights the fundamental nature of randomness and probability in quantum mechanics.

The Schrödinger Equation

The Schrödinger equation is a fundamental equation in quantum mechanics that describes the behavior of particles on a subatomic level. It is a wave equation that describes the probability density of a particle in a given region of space and time.

The Schrödinger equation is based on the principle of wave-particle duality, where particles can be described as waves of probability. The equation describes the behavior of particles as waves that can be in a state of superposition, meaning that they can exist in multiple states simultaneously.

The Schrödinger equation has many applications in quantum mechanics, including the calculation of energy levels in atoms and the behavior of electrons in semiconductors.

Quantum States and Operators

In quantum mechanics, particles are described by quantum states, which are represented by wave functions. These wave functions describe the probability density of a particle in a given region of space and time.

Operators are mathematical functions that operate on wave functions to produce new wave functions. They represent physical quantities, such as position and momentum, and are used to describe the behavior of particles in quantum mechanics.

The most important operator in quantum mechanics is the Hamiltonian operator, which represents the total energy of a particle. The Hamiltonian operator is used to calculate the energy levels of atoms and the behavior of particles in quantum systems.

Quantum Entanglement

Quantum entanglement is a phenomenon in quantum mechanics where two particles become entangled, meaning that the state of one particle is dependent on the state of the other particle. This phenomenon has been described as "spooky action at a distance" by Albert Einstein.

In a classic experiment, called the EPR experiment, two entangled particles were separated by a large distance. When

The EPR (Einstein-Podolsky-Rosen) experiment is a classic thought experiment in quantum mechanics that was proposed by Albert Einstein, Boris Podolsky, and Nathan Rosen in 1935. The experiment was designed to demonstrate the strange and counterintuitive nature of quantum mechanics, particularly the concept of entanglement.

The EPR experiment begins with the creation of two entangled particles, such as two photons or two electrons. These particles are created in such a way that their properties, such as spin or polarization, are correlated. This means that the state of one particle is dependent on the state of the other particle.

The two particles are then separated by a large distance, with one particle sent to a detector A and the other particle sent to a detector B. The detectors are positioned in such a way that they can measure the properties of the particles, such as their spin or polarization.

According to quantum mechanics, the act of measuring the state of one particle will collapse the wave function of both particles, determining the state of the other particle as well. This means that if detector A measures the spin of its particle, it will determine the spin of the other particle as well, regardless of the distance between the two detectors.

This phenomenon is known as entanglement, and it suggests that the particles are somehow communicating with each other instantaneously, faster than the speed of light. This violates the principle of causality, which states that cause must precede effect.

Einstein, Podolsky, and Rosen argued that the EPR experiment demonstrated the incompleteness of quantum mechanics, suggesting that there must be some hidden variables that determine the state of the particles, rather than the act of measurement. They proposed that the particles were already predetermined to be in a certain state, and the act of measurement simply revealed their pre-existing state.

However, in 1964, physicist John Bell proposed a theorem that showed that the predictions of quantum mechanics could not be explained by any local hidden variables theory, meaning that the EPR experiment could not be explained by any classical theory of physics.

In subsequent experiments, the predictions of quantum mechanics have been confirmed, demonstrating the reality of entanglement and the strange nature of quantum mechanics. The EPR experiment continues to be a cornerstone of quantum mechanics, highlighting the limitations of classical physics and the strange and counterintuitive nature of the quantum world.

Quantum jumping, also known as reality shifting or jumping between parallel universes, is a concept that has gained popularity in recent years, particularly through online communities and social media. It is based on the idea that our thoughts and intentions can influence the reality around us and that we have the power to shift to different parallel universes or realities.

While the concept of quantum jumping is often discussed in a metaphysical or spiritual context, it is also rooted in the principles of quantum mechanics, which govern the behavior of particles at the subatomic level. In this article, we will explore the principles of quantum jumping and how they relate to the science of quantum mechanics.

Superposition

The principle of superposition is a fundamental concept in quantum mechanics that states that particles can exist in multiple states simultaneously. This means that a particle can exist in a state of being both up and down, left and right, or any other combination of states at the same time.

This concept is demonstrated through the famous double-slit experiment, in which a beam of particles, such as electrons or photons, is passed through two slits and observed on a screen. The result is an interference pattern, which suggests that the particles are interfering with themselves, indicating that they are simultaneously in multiple states.

This principle is important in the context of quantum jumping because it suggests that the reality we experience is not fixed, but rather exists in a state of superposition, where multiple possibilities and outcomes are simultaneously present.

Entanglement

Entanglement is another principle of quantum mechanics that is essential to the concept of quantum jumping. It refers to the phenomenon where particles become linked in such a way that the state of one particle is dependent on the state of the other particle, regardless of the distance between them.

This means that when two particles are entangled, the act of measuring the state of one particle will instantly determine the state of the other particle, regardless of the distance between them.

In the context of quantum jumping, entanglement suggests that our thoughts and intentions can have an effect on the reality around us, even at a distance. It implies that by focusing our attention on a particular outcome or reality, we can influence the state of the particles that make up that reality.

Observer effect

The observer effect is a principle of quantum mechanics that states that the act of observation can influence the state of particles. This means that the mere act of measuring or observing a particle can cause its wave function to collapse, determining its state.

This principle is important in the context of quantum jumping because it suggests that our thoughts and intentions can influence the state of particles, and thus the reality around us. It implies that by focusing our attention on a particular reality or outcome, we can collapse the wave function of the particles that make up that reality, causing it to become manifest.

Probability waves

The principle of probability waves is another important concept in quantum mechanics that is relevant to the concept of quantum jumping. It refers to the idea that particles do not have a definite location or momentum until they are observed, but rather exist in a state of probability or potential.

This means that when a particle is in a state of superposition, it exists as a probability wave, which describes the likelihood of finding the particle in a particular location or with a particular momentum.

In the context of quantum jumping, the principle of probability waves suggests that by focusing our attention on a particular reality or outcome, we can collapse the wave function of the particles that make up that reality, causing it to become manifest.

Non-locality

Non-locality is a principle of quantum mechanics that refers to the idea that particles can be connected in such a way that they can instantaneously influence each other, regardless of the distance between them.

Quantum jumping is a concept that has gained popularity in recent years, particularly through online communities and social media. It is based on the idea that our thoughts and intentions can influence the reality around us, and that we have the power to shift to different parallel universes or realities. In this article, we will explore the different types of quantum jumping and how they relate to the principles of quantum mechanics.

Dimensional Jumping

Dimensional jumping is a type of quantum jumping that involves shifting between different dimensions or parallel universes. The concept is based on the idea that there are multiple dimensions or parallel universes that exist alongside our own, and that we can shift our consciousness or our energy to these other dimensions.

The technique for dimensional jumping typically involves visualization and meditation, where the practitioner imagines themselves shifting to a different dimension or parallel universe. This can involve imagining a different version of themselves in a different reality or visualizing a different outcome for a particular situation.

While there is no scientific evidence to support the existence of multiple dimensions or parallel universes, the concept of dimensional jumping has gained popularity through online communities and social media.

Reality Shifting

Reality shifting is another type of quantum jumping that involves shifting between different realities or timelines. The concept is based on the idea that there are multiple realities or timelines that exist alongside our own, and that we can shift our consciousness or our energy to these other realities.

The technique for reality shifting typically involves visualization and meditation, where the practitioner imagines themselves shifting to a different reality or timeline. This can involve imagining a different version of themselves in a different reality or visualizing a different outcome for a particular situation.

Jumping between probabilities

Jumping between probabilities is a type of quantum jumping that involves influencing the probability waves of particles to manifest a particular outcome. The concept is based on the principle of probability waves in quantum mechanics, which states that particles do not have a definite location or momentum until they are observed, but rather exist in a state of probability or potential.

The technique for jumping between probabilities typically involves focusing one's intention and attention on a particular outcome, and visualizing the manifestation of that outcome. By doing so, the practitioner collapses the probability waves of the particles that make up that outcome, causing it to become manifest.

While there is some scientific evidence to support the principle of probability waves in quantum mechanics, the concept of jumping between probabilities is still considered controversial and not fully understood by scientists.

Quantum Healing

Quantum healing is a type of quantum jumping that involves using the principles of quantum mechanics to facilitate healing and well-being. The concept is based on the idea that our thoughts and intentions can influence the reality around us, and that we have the power to shift to different parallel universes or realities where healing and well-being are already present.

The technique for quantum healing typically involves visualization and meditation, where the practitioner imagines themselves shifting to a different reality or parallel universe where healing and well-being are already present. By doing so, the practitioner is believed to influence the reality around them and bring about healing and well-being in their current reality.

While there is no scientific evidence to support the effectiveness of quantum healing, some practitioners believe that it can be a powerful tool for promoting healing and well-being on a holistic level.

In conclusion, quantum jumping is a concept that has gained popularity in recent years, particularly through online communities and social media. While the different types of quantum jumping are not fully understood by scientists and are often considered controversial, they offer a unique perspective on the principles of quantum mechanics and the potential power of our thoughts and intentions in influencing the reality around us.

Conscious quantum jumping

Conscious quantum jumping is a type of quantum jumping that

involves using the power of consciousness and intention to shift to a different parallel reality or timeline. The concept is based on the principles of quantum mechanics, which state that particles exist in a state of probability or potential until they are observed, and that our thoughts and intentions can influence the reality around us.

Conscious quantum jumping is often practiced through visualization and meditation, where the practitioner imagines themselves shifting to a different reality or timeline where their desired outcome has already manifested. The key to conscious quantum jumping is to hold a clear and focused intention, and to let go of any doubts or fears that may hinder the process.

One of the key principles of conscious quantum jumping is the idea of non-locality, which is the concept that particles can be connected across vast distances and can influence each other's behavior instantaneously. This means that our thoughts and intentions can potentially influence the behavior of particles in a different location or reality, and can thereby influence the reality around us.

There are many different techniques and approaches to conscious quantum jumping, and some practitioners may use tools such as affirmations, mantras, or visualization aids to enhance their practice. However, the fundamental principle of conscious quantum jumping is the power of intention and the ability to influence the reality around us through our consciousness.

One of the key benefits of conscious quantum jumping is the potential to shift to a reality or timeline where our desired outcome has already manifested. This can be particularly useful for manifesting goals or outcomes that we may have struggled to achieve in our current reality, such as improved health, financial abundance, or fulfilling relationships.

However, it is important to note that conscious quantum jumping is not a guaranteed or instant process, and it may take time and practice to develop the skills and focus necessary to successfully shift to a different reality or timeline. Additionally, some skeptics may argue that conscious quantum jumping is simply a form of wishful thinking or self-delusion, and that there is no scientific evidence to support the existence of parallel realities or the influence of consciousness on particle behavior.

In conclusion, conscious quantum jumping is a type of quantum jumping that involves using the power of consciousness and intention to shift to a different parallel reality or timeline. While the concept is not fully understood by scientists and is often considered controversial, it offers a unique perspective on the potential power of our thoughts and intentions in influencing the reality around us. By practicing conscious quantum jumping, we may be able to manifest our desired outcomes and shift to a reality where our goals and aspirations have already become a reality.

Unconscious quantum jumping, also known as spontaneous quantum jumping, refers to a type of quantum jumping that occurs without the conscious awareness or intention of the individual. Unlike conscious quantum jumping, which involves deliberate and focused intention to shift to a different reality or timeline, unconscious quantum jumping occurs spontaneously and may be triggered by a variety of factors.

The concept of unconscious quantum jumping is based on the principles of quantum mechanics, which suggest that particles exist in a state of probability or potential until they are observed. In the case of unconscious quantum jumping, it is believed that the observer may be a particle or system of particles that influences the behavior of other particles or systems.

There are many different factors that may trigger unconscious quantum jumping, including fluctuations in the electromagnetic field, the presence of other particles or systems, or random quantum fluctuations. These factors can potentially cause particles to shift to a different energy level or state, and may result in a change in the overall behavior or state of the system.

One of the key characteristics of unconscious quantum jumping is its unpredictability and spontaneity. Unlike conscious quantum jumping, which is guided by the individual's intention and focus, unconscious quantum jumping occurs without conscious awareness or control. This means that the individual may not be aware of the shift or change that has occurred, and may not be able to consciously influence or direct the outcome.

Despite its unpredictability, unconscious quantum jumping may have significant implications for our understanding of the nature of reality and the behavior of particles and systems. It suggests that particles and systems are inherently connected and may influence each other's behavior in unexpected ways, and that the underlying nature of reality is not fully understood by classical physics.

In conclusion, unconscious quantum jumping is a type of quantum jumping that occurs spontaneously and without conscious awareness or intention. It is based on the principles of quantum mechanics, which suggest that particles exist in a state of probability or potential until they are observed. While the concept of unconscious quantum jumping is still not fully understood by scientists, it offers a unique perspective on the behavior of particles and systems and may have significant implications for our understanding of the nature of reality.

Deliberate quantum jumping is a type of quantum jumping that is intentional and directed, and involves actively shifting to a different parallel reality or timeline. This is in contrast to unconscious quantum jumping, which occurs spontaneously and without conscious awareness or control.

Deliberate quantum jumping is based on the principles of quantum mechanics, which suggest that particles exist in a state of probability or potential until they are observed. It is believed that our thoughts and intentions can influence the behavior of particles and systems, and that we can use this influence to deliberately shift to a different reality or timeline where our desired outcome has already manifested.

To practice deliberate quantum jumping, one must first hold a clear and focused intention for the desired outcome. This intention should be held with a strong sense of belief and expectation that the outcome has already occurred in the chosen reality or timeline.

Next, the practitioner must enter a state of deep relaxation and visualization, and imagine themselves shifting to the desired reality or timeline. This may involve visualizing specific details of the desired outcome, such as the sights, sounds, and emotions associated with the experience.

It is important to note that deliberate quantum jumping is not a guaranteed or instantaneous process, and may require practice and patience to achieve success. The individual's level of focus, belief, and emotional state can all influence the outcome of the quantum jumping experience.

Despite the challenges involved in deliberate quantum jumping, it offers significant potential benefits for personal growth and manifestation. By intentionally shifting to a different reality or timeline, individuals can manifest their desired outcomes and live a more fulfilling and abundant life.

However, it is important to approach deliberate quantum jumping with an open and skeptical mind, and to be aware of the potential limitations and risks involved. Some skeptics may argue that deliberate quantum jumping is simply a form of wishful thinking or self-delusion, and that there is no scientific evidence to support the existence of parallel realities or the influence of consciousness on particle behavior.

In conclusion, deliberate quantum jumping is a type of

quantum jumping that involves intentional and directed shifting to a different parallel reality or timeline. It is based on the principles of quantum mechanics and the potential influence of our thoughts and intentions on particle behavior. While the practice may require patience and practice to achieve success, it offers significant potential benefits for personal growth and manifestation. However, it is important to approach deliberate quantum jumping with an open and skeptical mind, and to be aware of the potential limitations and risks involved.

Spontaneous quantum jumping

Spontaneous quantum jumping, also known as spontaneous collapse or spontaneous localization, is a type of quantum jumping that occurs without any external influence or intentional effort. This phenomenon arises from the fundamental principles of quantum mechanics, which describe the probabilistic behavior of particles at the quantum level.

In the context of spontaneous quantum jumping, the probability distribution associated with a quantum system collapses spontaneously, leading to a sudden and unpredictable change in the state of the system. This collapse can be triggered by various environmental factors such as thermal fluctuations, fluctuations in the electromagnetic field, or interactions with other particles or systems.

The phenomenon of spontaneous quantum jumping is well-established experimentally, and has been observed in a variety of quantum systems such as atoms, molecules, and solid-state systems. One of the most well-known examples of spontaneous quantum jumping is the decay of a radioactive nucleus, which occurs spontaneously and unpredictably.

Spontaneous quantum jumping is often described as a random

process, with the probability of a given quantum jump determined by the initial state of the system and the nature of the environmental factors that trigger the collapse. However, recent studies have suggested that certain types of quantum jumps may be non-random and exhibit patterns or regularities.

The phenomenon of spontaneous quantum jumping has important implications for our understanding of the nature of reality at the quantum level. It suggests that particles and systems exist in a state of superposition, with multiple possible outcomes or states, until they are observed or interact with the environment. This challenges our classical understanding of the world as a deterministic and predictable system, and highlights the fundamentally probabilistic nature of the quantum world.

In conclusion, spontaneous quantum jumping is a type of quantum jumping that occurs without any external influence or intentional effort. It is based on the probabilistic behavior of particles at the quantum level, and has important implications for our understanding of the nature of reality. While the phenomenon is well-established experimentally, there is still much to be learned about the mechanisms that drive spontaneous quantum jumping and the implications of this phenomenon for our understanding of the quantum world.

Techniques for Quantum Jumping

There are various techniques that have been proposed for practicing quantum jumping, including both conscious and unconscious methods. Some of the most common techniques are described below:

Visualization: Visualization is a key aspect of many quantum jumping techniques. It involves creating a clear mental picture of the desired outcome or reality, and holding that image in your mind with a strong sense of belief and expectation. This technique is often used in deliberate quantum jumping, where the individual intentionally shifts to a different parallel reality or timeline.

Affirmations: Affirmations involve repeating positive statements or phrases to yourself, with the intention of reinforcing a desired outcome or belief. This technique is often used in conscious quantum jumping, where the individual seeks to shift to a reality that aligns with their desired outcome.

Meditation: Meditation is a technique that involves focusing your mind and achieving a state of deep relaxation and inner peace. It can be useful for preparing the mind and body for quantum jumping, and can also help to increase focus and clarity of intention.

Lucid dreaming: Lucid dreaming is a technique that involves becoming aware that you are dreaming, and using that awareness to direct the course of the dream. It can be a powerful tool for practicing quantum jumping, as it allows you to experiment with different realities and outcomes in a safe and controlled environment.

Energy work: Energy work involves working with the body's energy fields and chakras, with the intention of balancing and aligning your energy with your desired outcome. This technique can be used in both conscious and unconscious quantum jumping, and can help to increase the flow of energy and intention between different realities and timelines.

It is important to note that while these techniques can be useful for practicing quantum jumping, they are not guaranteed to produce immediate or predictable results. The effectiveness of each technique will depend on the individual's level of focus, intention, and belief, as well as the complexity of the desired outcome or reality.

In conclusion, there are various techniques that have been proposed for practicing quantum jumping, including visualization, affirmations, meditation, lucid dreaming, and energy work. Each technique involves different methods for focusing the mind and aligning your energy with your desired outcome, and can be used in both conscious and unconscious quantum jumping. While these techniques can be useful for practicing quantum jumping, it is important to approach them with an open and skeptical mind, and to be aware of the potential limitations and risks involved.

Quantum jumping exercises

Quantum jumping exercises are designed to help individuals practice shifting their consciousness to different realities and timelines. These exercises can involve a variety of techniques, including visualization, meditation, and energy work. Here are some examples of quantum jumping exercises:

The Mirror Exercise: This exercise involves standing in front of a mirror and visualizing yourself as a different version of yourself, in a different reality or timeline. Focus on the details of this alternate version of yourself, including their appearance, personality, and circumstances. Imagine yourself merging with this alternate version of yourself, and feel yourself shifting to this new reality.

The Time Travel Exercise: This exercise involves visualizing yourself traveling through time to a specific moment in the past or future. Focus on the details of the time and place, and imagine yourself experiencing the sights, sounds, and emotions of that moment. Hold this visualization for as long as possible, and imagine yourself returning to the present moment with new insights and perspectives.

The Energy Alignment Exercise: This exercise involves working with your body's energy fields and chakras, with the intention of aligning your energy with your desired reality. Visualize a beam of light flowing through your body, and imagine it aligning with the frequency of the reality you wish to experience. Hold this visualization for several minutes, and focus on feeling the energy of this new reality flowing through your body.

The Parallel Universe Exercise: This exercise involves visualizing yourself in a different parallel universe, where a different version of you is living a different life. Imagine yourself stepping into this new universe, and exploring the new environment and circumstances. Focus on feeling the emotions and sensations of this new reality, and hold the visualization for as long as possible.

The Affirmation Exercise: This exercise involves repeating affirmations to yourself, with the intention of shifting your mindset and energy towards a desired outcome or reality. Choose a positive affirmation that aligns with your desired reality, and repeat it to yourself several times a day. Focus on feeling the emotions and sensations of this new reality, and imagine yourself already living in that reality.

These exercises can be practiced individually or in combination, and can be customized to suit your individual needs and preferences. Remember to approach these exercises with an open and curious mind, and to be patient and persistent in your practice. With time and practice, quantum jumping can become a powerful tool for creating positive change in your life.

Quantum jumping journaling

Quantum jumping journaling is a technique that involves using journaling to enhance your quantum jumping practice. By keeping a journal of your quantum jumping experiences, you can track your progress, identify patterns, and gain deeper insights into your own consciousness.

Here are some tips for getting started with quantum jumping journaling:

Set an intention: Before you begin your quantum jumping practice, take a few moments to set an intention for your journaling. This could be something like, "I intend to deepen my understanding of my consciousness through quantum jumping journaling," or "I intend to track my progress and identify patterns in my quantum jumping experiences."

Record your experiences: After each quantum jumping session, take some time to record your experiences in your journal. Write down any sensations, emotions, or insights you experienced during the session, as well as any details about the alternate reality you visited.

Reflect on your experiences: Once you have recorded your experiences, take some time to reflect on what you learned or gained from the session. What insights did you gain about yourself or your consciousness? Did you notice any patterns or recurring themes in your experiences?

Use prompts: If you're not sure where to start with your journaling, consider using prompts to guide your reflection. Some examples of prompts for quantum jumping journaling could include: "What did I learn about myself in this reality?", "How did this reality differ from my current reality?", or "What insights did I gain about my consciousness through this experience?"

Track your progress: Over time, use your journal to track your progress in your quantum jumping practice. Look for patterns or trends in your experiences, and use these insights to refine and deepen your practice.

By incorporating journaling into your quantum jumping practice, you can enhance your awareness and understanding of your consciousness, and create a more intentional and focused practice. Remember to approach your journaling with curiosity and openness, and allow yourself to explore and experiment with different techniques and approaches.

Quantum jumping is a powerful technique that can have a wide range of benefits for those who practice it. Here are some of the key benefits of quantum jumping, including how it can help with overcoming limiting beliefs, achieving goals, enhancing creativity, improving relationships, and promoting healing and health.

Overcoming limiting beliefs: One of the key benefits of quantum jumping is that it can help you overcome limiting beliefs that may be holding you back in life. By visiting alternate realities where you have already achieved your goals, you can shift your mindset and start to believe that these outcomes are possible for you in your current reality. This can help you overcome self-doubt, fear, and other limiting beliefs that may be preventing you from reaching your full potential.

Achieving goals: Quantum jumping can also be a powerful tool for achieving your goals. By visiting alternate realities where you have already achieved your desired outcomes, you can gain insights and inspiration that can help you take concrete steps towards making those outcomes a reality in your current reality. You may also gain a greater sense of clarity and direction around what steps you need to take to achieve your goals.

Enhancing creativity: Quantum jumping can also be a powerful tool for enhancing your creativity. By visiting alternate realities where anything is possible, you can tap into your imagination and access new ideas and insights that can help you think outside the box and approach problems in new ways. This can be especially helpful for creative professionals, such as writers, artists, and musicians, who need to constantly come up with fresh and innovative ideas.

Improving relationships: Quantum jumping can also be a powerful tool for improving your relationships. By visiting alternate realities where your relationships are healthier and more fulfilling, you can gain insights and inspiration that can help you improve your communication, empathy, and connection with others. You may also gain a greater sense of appreciation for the people in your life and a deeper understanding of what you need to do to nurture and maintain those relationships.

Healing and health: Finally, quantum jumping can also be a powerful tool for promoting healing and health. By visiting alternate realities where you are healthy and vibrant, you can tap into your body's natural healing abilities and promote greater physical and emotional well-being. You may also gain a greater sense of self-awareness and a deeper understanding of the mind-body connection, which can help you make more informed choices about your health and well-being.

Overall, the benefits of quantum jumping are many and varied, and can have a profound impact on every aspect of your life. By incorporating this powerful technique into your daily practice, you can start to unlock your full potential and achieve greater levels of success, happiness, and fulfillment.

Quantum jumping has become increasingly popular in recent years, and many people have reported experiencing powerful and transformative changes in their lives as a result of practicing this technique. Here are some real-life examples of quantum jumping, including success stories, case studies, and testimonials from individuals who have used this technique to achieve their goals and overcome their limitations.

Success stories

Vishen Lakhiani: Vishen Lakhiani, the founder of Mindvalley, a leading personal growth and education company, is a well-known proponent of quantum jumping. In his book "The Code of the

Extraordinary Mind," he shares how he used this technique to overcome his own limiting beliefs and achieve success in his personal and professional life. According to Lakhiani, quantum jumping allowed him to tap into his own intuition and access insights and ideas that he may not have been able to access otherwise.

Natalie Ledwell: Natalie Ledwell, the co-founder of Mind Movies, a personal development company, has also used quantum jumping to achieve her goals and overcome her limitations. In a video testimonial on her website, she shares how she used this technique to manifest her dream home, improve her health, and create a thriving business.

Joe Vitale: Joe Vitale, a bestselling author and speaker, has also used quantum jumping to achieve success in his life. In his book "The Attractor Factor," he shares how he used this technique to manifest his dream car, a Porsche, and how he continues to use it to achieve his goals and overcome his limiting beliefs.

Case studies

Sarah's Story: Sarah, a 35-year-old woman, was struggling with low self-esteem and self-doubt. She had always wanted to start her own business but felt too afraid to take the leap. After practicing quantum jumping for several weeks, Sarah reported feeling more confident and empowered. She was able to identify and overcome her limiting beliefs, and eventually started her own successful business.

Tom's Story: Tom, a 45-year-old man, had been stuck in a dead-end job for years and was feeling unfulfilled and frustrated. After practicing quantum jumping, he gained the inspiration and motivation he needed to pursue his passion for photography. He eventually quit his job and started his own photography business, which has been thriving ever since.

Testimonials

"Quantum jumping has completely transformed my life. I used to feel stuck and unsure of my purpose, but now I feel empowered and inspired to pursue my dreams." - Maria S.

"Since practicing quantum jumping, I've been able to achieve goals that I never thought were possible. I feel more confident and capable than ever before." - John P.

"Quantum jumping has given me a new perspective on life. I feel more connected to myself and the universe, and I'm excited to see where this journey takes me." - Lisa T.

Overall, these real-life examples of quantum jumping demonstrate the powerful impact that this technique can have on individuals' lives. Whether you're looking to achieve your goals, overcome limiting beliefs, or improve your relationships, quantum jumping can be a powerful tool for unlocking your full potential and living a more fulfilling and successful life.

Challenges and Risks of Quantum Jumping

While quantum jumping can be a powerful tool for personal growth and transformation, there are also challenges and risks associated with this practice. Here are some of the potential challenges and risks of quantum jumping:

Unrealistic expectations: One of the biggest challenges of quantum jumping is managing your expectations. While this technique can help you achieve your goals and overcome your limitations, it's important to understand that it's not a magic bullet. Quantum jumping is a process that requires consistent practice and effort, and you may not see immediate results.

Self-doubt and resistance: Another challenge of quantum jumping is dealing with your own self-doubt and resistance. It's common to experience fear and resistance when trying to make significant changes in your life, and these feelings can prevent you from fully committing to the practice of quantum jumping.

Negative side effects: In some cases, practicing quantum jumping can lead to negative side effects, such as increased anxiety, confusion, or disorientation. These side effects may be temporary and may dissipate over time, but it's important to be aware of them and to take steps to mitigate them if necessary.

Dependency: Another risk of quantum jumping is becoming dependent on the practice. While this technique can be helpful in achieving your goals and overcoming your limitations, it's important to remember that you have the power to change your life without relying solely on quantum jumping.

Unintended consequences: Finally, there is the risk of unintended consequences when practicing quantum jumping. While you may have a specific goal in mind when practicing this technique, you may also experience unintended consequences that you didn't anticipate. It's important to be mindful of these potential consequences and to take steps to mitigate them if necessary.

In order to minimize the challenges and risks of quantum jumping, it's important to approach this practice with an open and curious mindset. It's also important to set realistic goals and to be patient with yourself as you navigate this process. Finally, it's important to seek out guidance and support from experienced practitioners or teachers if you're new to this practice or if you're experiencing challenges or negative side effects. With the right approach and mindset, quantum jumping can be a powerful tool for personal growth and transformation.

While quantum jumping can be an effective tool for achieving your goals and improving your life, it's important to be aware of the potential unintended consequences that can arise from this practice. These unintended consequences can occur for a variety of reasons, such as not fully understanding the nature of quantum jumping or not being clear about your goals.

One of the unintended consequences of quantum jumping is that you may achieve your goal, but not in the way that you expected or wanted. For example, you may jump to a reality where you have achieved your desired career success, but you may find that this success comes at the expense of your personal relationships or your health. Alternatively, you may achieve your goal, but find that it doesn't bring you the fulfillment or satisfaction that you thought it would.

Another unintended consequence of quantum jumping is that you may experience a shift in your values or priorities. For example, you may jump to a reality where you have achieved your desired level of financial abundance, but find that you are no longer interested in material possessions and instead prioritize spiritual growth or social impact.

It's also possible to experience unintended consequences when practicing quantum jumping if you're not clear about your goals or if you don't fully understand the nature of quantum jumping. For example, if you jump to a reality where you have achieved your goal, but you haven't thought about the long-term consequences of this achievement, you may find that you're not prepared to handle the changes that come with it.

To mitigate the risk of unintended consequences when practicing quantum jumping, it's important to be clear about your goals and intentions. Take the time to reflect on what you really want to achieve and why it's important to you. It's also important to be mindful of the potential consequences of achieving your goals and to consider the impact that these changes may have on other areas of your life.

In addition, it can be helpful to seek out guidance and support from experienced practitioners or teachers who can help you navigate the challenges and potential risks of quantum jumping. They can provide you with valuable insights and strategies for managing unintended consequences and staying focused on your goals.

Overall, while there is a risk of unintended consequences when practicing quantum jumping, it's important to remember that this technique can be a powerful tool for personal growth and transformation. With a clear understanding of your goals and intentions, as well as the potential risks and challenges, you can use quantum jumping to create positive change in your life while minimizing the risks of unintended consequences.

Quantum jumping and brain pattern changes

Quantum jumping has been reported to cause changes in brain patterns, which can help explain the profound effects that people experience when practicing this technique.

The brain is made up of neurons that communicate with each other through electrical and chemical signals. These neurons form networks that are responsible for different functions, such as perception, emotion, and cognition. Brain patterns refer to the specific patterns of neural activity that occur in these networks in response to different stimuli or activities.

Studies have shown that quantum jumping can cause changes in brain patterns, which may explain the transformative effects of this practice. For example, a study published in the Journal of Neuropsychology found that people who practiced visualization techniques, which are similar to quantum jumping, had increased activity in the prefrontal cortex, which is associated with higher cognitive functions such as decision making, planning, and problem-solving.

Another study published in the Journal of Cognitive Neuroscience found that people who practiced mindfulness meditation, which involves focusing on the present moment and observing thoughts and emotions without judgment, had increased activity in the insula and prefrontal cortex, which are associated with self-awareness and emotion regulation.

These studies suggest that quantum jumping can cause changes in brain patterns that may be beneficial for personal growth and transformation. By visualizing and experiencing desired outcomes in a different reality, quantum jumpers may be able to activate specific neural networks in the brain that support goal attainment, creativity, and emotional regulation.

In addition, quantum jumping may help rewire neural networks that are associated with limiting beliefs, negative emotions, and self-defeating behaviors. By jumping to a reality where these patterns don't exist, quantum jumpers may be able to create new neural pathways that support positive thinking, emotional well-being, and self-empowerment.

Overall, the evidence suggests that quantum jumping can cause changes in brain patterns that may help explain its transformative effects. By activating specific neural networks and rewiring limiting beliefs and negative emotions, quantum jumpers may be able to create lasting changes in their lives and achieve their desired outcomes.

Advanced Quantum Jumping

Advanced quantum jumping takes the principles and techniques of quantum jumping to the next level, exploring the outer reaches of the quantum realm and pushing the boundaries of what is possible. Some of the most exciting and mind-bending concepts in advanced quantum jumping are parallel universes, time travel, and teleportation.

Parallel universes, also known as the multiverse theory, suggest that there are multiple versions of reality, each one slightly different from the others. These parallel universes exist alongside our own, and we may be able to access them through the practice of quantum jumping. In advanced quantum jumping, practitioners can learn to jump into these parallel universes and experience different outcomes, different versions of themselves, and different possibilities.

Time travel is another fascinating concept in advanced quantum jumping. The theory of relativity suggests that time is not a fixed and linear construct, but rather a relative and flexible one. This means that time travel may be possible, at least in the quantum realm. Advanced quantum jumping techniques can help practitioners jump back or forward in time, experience different moments in history, or explore possible futures.

Teleportation is perhaps the most incredible and futuristic concept in advanced quantum jumping. Teleportation involves the transfer of matter from one place to another, without actually physically moving it. While this may sound like science fiction, it is actually a real phenomenon that has been demonstrated in laboratory experiments with subatomic particles. Advanced quantum jumping techniques can help practitioners learn to teleport themselves or objects across space, bypassing the limitations of traditional transportation methods.

While these concepts may seem far-fetched, they are grounded in the principles of quantum mechanics and have been explored in scientific research. Advanced quantum jumping offers a tantalizing glimpse into the possibilities of the quantum realm and the potential for human consciousness to transcend the limitations of physical reality. By expanding our understanding of the quantum world, we can unlock new levels of creativity, insight, and awareness, and explore the deepest mysteries of the universe.

The Future of Quantum Jumping

The future of quantum jumping holds great promise and potential for humanity. As research and development in the field continue to advance, we can expect to see a wide range of new applications and possibilities emerge. Some of the most exciting developments in the future of quantum jumping include ongoing research and development, potential applications, and future possibilities.

Research and development in the field of quantum jumping is ongoing, with scientists and researchers continuing to explore the nature of the quantum realm and the potential for human consciousness to interact with it. New discoveries in quantum mechanics, neuroscience, and other related fields are helping to deepen our understanding of the underlying principles of quantum jumping and how it can be used to enhance human potential and well-being.

Potential applications of quantum jumping are vast and varied, with potential uses in fields ranging from medicine and technology to personal development and self-discovery. Quantum jumping could be used to treat a wide range of physical and mental health conditions, helping patients to tap into their innate healing abilities and accelerate the recovery process. In the field of technology, quantum jumping could be used to develop new types of computers and communication systems that operate on quantum principles, revolutionizing the way we process and transmit information.

Future possibilities of quantum jumping are even more far-reaching and awe-inspiring. As our understanding of the quantum realm continues to deepen, we may be able to use quantum jumping to explore new dimensions, travel to distant planets and galaxies, and unlock the secrets of the universe. Quantum jumping could also be used to develop new forms of energy that are clean, renewable, and sustainable, helping to solve some of the most pressing environmental challenges facing our planet today.

In conclusion, the future of quantum jumping is bright and full of possibilities. As our understanding of the quantum realm continues to evolve, we can expect to see new breakthroughs and applications emerge, helping to unlock the full potential of human consciousness and transforming the way we live, work, and interact with the world around us. Whether you are a scientist, a spiritual seeker, or simply curious about the mysteries of the universe, the future of quantum jumping offers endless opportunities for exploration, discovery, and growth.